This edition published in 1991 by Rainbow Books,
Elsley House, 24–30 Great Titchfield Street, London, W1P 7AD.

Originally published in 1989 as *Animal Life Stories: The Mouse* by Kingfisher Books.

ISBN 1 871745 27 6

Black and white illustrations by Jean Colville.
End papers by Maurice Pledger.

Printed and bound in Spain.

EYE-VIEW LIBRARY

THE MOUSE

By Angela Royston

Illustrated by Maurice Pledger

RAINBOW
·BOOKS·

Clouds cover the moon and darkness hides the mouse as she feeds on grass and insects. Then a wind blows the clouds away and a weasel catches a glimpse of her in the moonlight. She quickly hides behind a clump of grass. The weasel sees a vole and chases it, and the mouse hurries home to her nest.

Her babies have not yet left the nest. They feed from her milk and explore the tunnels in their burrow. Sometimes they peep from the entrance holes at the world outside.

One day they follow their mother out a little way, but they are frightened. The woods are full of strange smells and noises. An owl hoots and they rush back to the nest.

The young mice soon learn to run quickly and quietly. As they nibble their food they listen out for foxes, weasels and other dangerous animals.

When they are three weeks old the young mice leave their mother and make their own burrows. One of them digs out a hole beneath a fallen tree trunk where it will be safe and warm.

The mother stays in the nest on her own, sleeping during the day and finding food at night. While she is nibbling some wild strawberries one night she hears a scratching noise behind her. It is a male mouse. The two animals sniff each other carefully and then they mate. Suddenly an owl swoops by and they scamper back to their own burrows.

Just three weeks and three days later, the mother mouse has a new litter of pink, squeaking babies. She often leaves them while she hunts for food.

Very early one morning she is hurrying back to her nest when a terrible sight stops her. A badger has broken into her burrow and destroyed her nest.

The mother mouse is alone again. She hides in an old log by the side of a stream and that night she digs a new burrow underneath it. The next evening, while she is nibbling some nuts, she sees her friend the male mouse and they mate again. Soon she has a new litter of babies.

It is autumn and the mother mouse is busy storing up nuts and berries to feed her growing babies. One night there is a big storm. The rain pours down for many hours and the stream becomes swollen with rushing water.

Suddenly a great crash of thunder frightens all the animals in the wood. A young deer runs blindly past. Her hoof knocks the log and pushes it into the stream. Most of the mouse's nest is dragged into the water with it.

Squeaking with fear, the young mice cling to the grassy edge of the bank. One by one their mother pulls them out and drags them to safety.

That night she digs a new burrow. The woods are full of nuts and berries now and she gathers a big store of them to feed herself and her family. Soon the young mice leave her nest and make burrows of their own.

Winter comes and the mouse spends most of her time sleeping. One day she wakes up to find a squirrel eating some of her store of nuts. Luckily the mouse does not need all the nuts because she ate so much in autumn.